PROCESS INTELLIGENCE

The Missing Link Between Agile, DevOps, and AI

V. Quan Lee & J. Kay

From Intelligent Systems Press

An imprint dedicated to advancing system design, process intelligence, and modern delivery thinking

www.IntelligentSystemsPress.com

INTELLIGENT
SYSTEMS PRESS
IDEAS. SYSTEMS. INTELLIGENT IMPACT.

COPYRIGHT

COPYRIGHT PAGE

Published by Intelligent Systems Press

Intelligent Systems Press is an independent publishing imprint focused on system design, process intelligence, and modern delivery optimization.

ISBN: 979-8-9958725-0-4

First Edition

Editing by: Brian Loveless
Web Content Design by: Erik Victor Jones
Content Editing and Layout by: Emma Victoria Lee

For permissions, inquiries, or bulk orders, contact:

Customerservice@intelligentsystemspress.com

CONTENTS

TABLE OF CONTENTS

THE EVOLUTION OF INTELLIGENT PROCESS THINKING

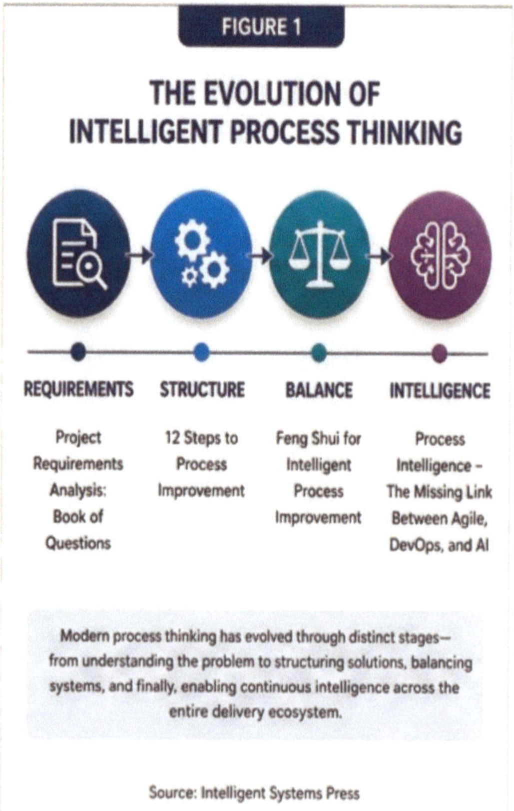

PREFACE

Most organizations believe they have a delivery problem.

They don't.

They have a system problem.

Across industries, companies have invested billions into Agile transformations, DevOps pipelines, and AI-driven capabilities—yet the same issues persist:

Delivery is inconsistent

Outcomes are unpredictable

Teams are misaligned

Performance does not scale

On the surface, everything appears modern.

Underneath, the system is fragmented.

This is the hidden reality of modern software delivery:

We have optimized parts—but never truly understood the system.

Agile improves adaptability. DevOps improves flow. AI accelerates execution.

But none of them explain how work actually moves across the system.

And that is where failure lives.

The Missing Layer

This book introduces Process Intelligence—the missing operating layer between Agile execution, DevOps delivery, and AI-driven systems.

Process Intelligence is not another framework. It is a shift in perspective:

From managing tasks → to understanding systems

From improving teams → to aligning flow

From measuring activity → to optimizing outcomes

It provides a way to see what has always been there—but rarely understood:

- How work truly flows.
- Where it breaks.
- Why it slows.
- And how to fix it—continuously.

Why This Matters Now

For decades, organizations have tried to solve delivery challenges by adding more:

THE PATTERN

More tools. More frameworks. More automation.

But complexity has increased faster than clarity.

And complexity without clarity creates fragility.

In today's environment—where systems are interconnected, fast-moving, and AI-augmented—the ability to understand flow is no longer optional. It is a competitive advantage.

What This Book Will Do

This book will not give you another methodology to implement.

It will give you something far more powerful:

A way to see your system clearly.

Because once a system is understood, it can be improved. Continuously. Intentionally. Intelligently.

Process Intelligence Manifesto

WE BELIEVE

Work does not fail because people are incapable—

It fails because systems are misunderstood.

Speed without alignment creates chaos.

Automation without clarity accelerates failure.

Optimization without visibility is guesswork.

We reject the idea that more tools solve system problems.

We believe that:

- Clarity creates alignment.
- Alignment creates performance.
- Performance creates continuous improvement.

Process Intelligence is not a methodology.
It is a shift in how systems are understood.
And once seen, it cannot be unseen.

The Category Definition

Process Intelligence is not an evolution of Agile or DevOps. It is the operating layer above them.

Existing Frameworks	**Process Intelligence**
Agile → optimizes how teams build	**Optimizes how the entire system performs**
DevOps → optimizes how systems deliver	✓ See system-level flow
AI → optimizes how execution scales	✓ Align cross-functional execution
	✓ Continuously optimize outcomes

EXECUTIVE INSIGHT

Organizations that adopt Process Intelligence will not move faster. They will move correctly. And in complex systems—correctness scales faster than speed.

CHAPTER 1

THE EVOLUTION OF SOFTWARE DELIVERY

FIGURE 2

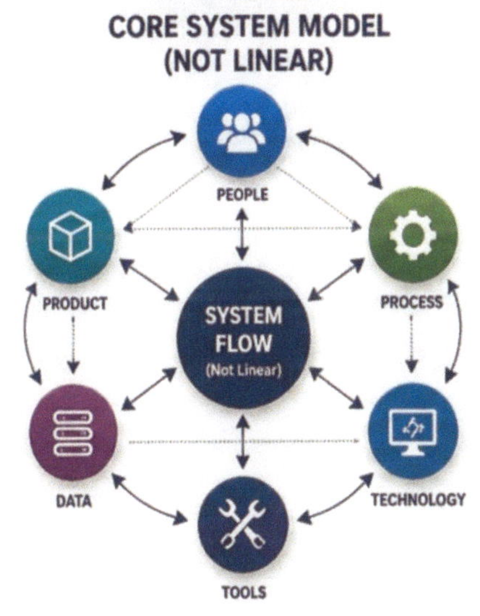

Source: Intelligent Systems Press

Software delivery has undergone multiple transformations over the past several decades. Each shift—from Waterfall to Agile, from Agile to DevOps, and now toward AI-enabled systems—has attempted to address limitations in how work is executed.

Yet despite these advancements, a consistent pattern remains: organizations continue to struggle with misalignment, inefficiency, and unpredictable outcomes.

CORE INSIGHT

The reason is not a lack of methodology. It is a lack of system-level understanding.

The Limitation of Linear Thinking

Traditional process models were designed around linear progression. Work was expected to move from one stage to the next in a predictable sequence.

This assumption held true in environments where requirements were stable and change was limited.

Modern systems do not operate this way.

Work flows through interconnected pathways. Dependencies, constraints, and feedback loops influence how progress occurs. A delay in one area creates ripple effects across the entire system.

Linear thinking simplifies reality—but it also obscures it.

CASE — THE FLOW CONSTRAINT

Consider a development team that increases its output by completing more work during each sprint. On the surface, this appears to be an improvement.

However, if downstream testing environments are constrained, the additional output creates a backlog. Work accumulates, delays increase, and system performance declines.

The issue is not the team's productivity. The issue is system flow.

CHAPTER SYNTHESIS

This chapter establishes a foundational shift:

- From viewing work as a sequence of steps
- To understanding it as a system of flow

Without this shift, organizations will continue to optimize isolated components while overall performance remains constrained.

System performance is determined by how work moves—not how much work is completed.

C H A P T E R 2

AGILE FRAMEWORKS IN REALITY

Agile transformed how organizations approach software delivery by introducing adaptability, iteration, and responsiveness to change. It replaced rigid, long-cycle planning with shorter feedback loops.

However, despite widespread adoption, many organizations fail to realize the full benefits of Agile.

THE ROOT CAUSE

The issue is not Agile itself. The issue is how it is applied.

Agile is often implemented as a set of ceremonies rather than a system of flow. Teams conduct stand-ups, plan sprints, and hold retrospectives—but underlying system constraints remain unchanged.

As a result, Agile improves activity—but not outcomes.

CASE — THE VELOCITY TRAP

A team improves sprint velocity over multiple cycles. Output increases.

- Testing queues grow
- Releases slow
- Defects increase

The team is performing. The system is failing.

Real-World Scenario

A product team increases its sprint velocity over multiple iterations. Story points completed per sprint steadily rise. However, release timelines remain inconsistent.

Because deployment pipelines, testing environments, and cross-team dependencies are not aligned with the increased output. The team has optimized locally. The system has not improved.

Agile enables teams to move faster—but without system alignment, increased speed leads to increased congestion.

EXECUTIVE INSIGHT

Agile fails when systems remain misaligned. To realize the full value of Agile, organizations must extend their focus beyond individual teams. They must understand how work flows across the entire system.

CHAPTER SYNTHESIS

Agile represents a significant advancement in adaptability. However, it does not solve the problem of system alignment.

Without visibility into how work flows across teams and processes, improvements remain localized. The next evolution requires expanding beyond team-level optimization to system-level understanding.

CHAPTER 3

DEVOPS AND CONTINUOUS DELIVERY

DevOps emerged as a response to one of the most persistent challenges in software delivery: the disconnect between development and operations. Where Agile improved how teams build software, DevOps aimed to improve how software is delivered.

It introduced shared ownership, automation, and continuous delivery pipelines designed to accelerate release cycles and reduce friction between teams.

THE PROBLEM

In many organizations, DevOps becomes overly focused on tooling. Pipelines are built. Automation is implemented. Deployment frequency increases.

Yet underlying inefficiencies remain. Automation does not fix broken systems. It amplifies them.

CASE — THE AUTOMATION ILLUSION

An organization invests heavily in building a robust CI/CD pipeline. Deployments become faster and more frequent. However, failure rates increase. Rollbacks become more common. Production issues begin to surface more frequently.

The system became faster, but not better. Speed without alignment creates instability.

EXECUTIVE INSIGHT

DevOps succeeds when ownership is shared—not transferred. True DevOps is not about handing work from development to operations more efficiently. It is about eliminating the boundaries between them.

Process Anti-Patterns

Most organizations do not fail because they lack solutions. They fail because they reinforce ineffective patterns.

COMMON ANTI-PATTERNS

Local Optimization

Teams improve their own performance, but system outcomes decline due to misalignment.

Automation Without Clarity

Processes are automated before they are understood, accelerating inefficiencies.

Metric Misalignment

Teams measure activity instead of outcomes, creating the illusion of progress.

Over-Engineering

Complex solutions are introduced to solve simple problems, increasing system friction.

Reactive Decision-Making

Problems are addressed only after they occur, rather than through proactive system understanding.

System Failure Patterns

Failure Pattern	Impact
The Velocity Illusion Teams increase output. System declines.	**The Metric Mirage** Metrics improve. Outcomes decline.
The Automation Trap Automation scales inefficiency.	**The Tool Dependency Spiral** More tools, less clarity.
The Alignment Drift Systems slowly fall out of sync.	***Core Insight: These patterns are caused by system design, not people.***

CHAPTER SYNTHESIS

DevOps represents a major advancement in enabling flow. However, flow without understanding is incomplete.

Automation increases speed—but without alignment and visibility, it also increases risk.

The next step is not more tools. It is deeper system awareness.

CHAPTER 4

PROCESS INTELLIGENCE

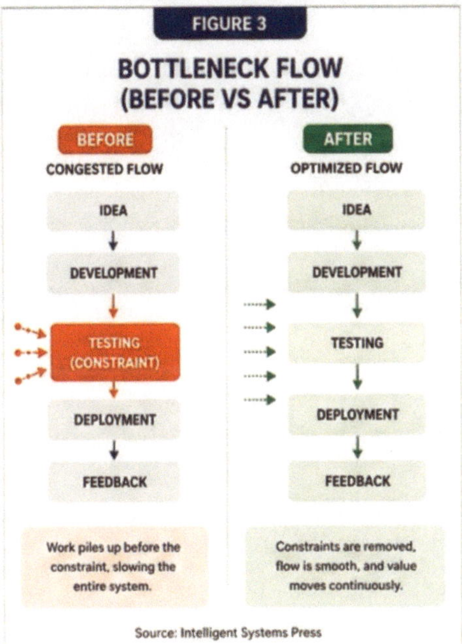

Process Intelligence represents a fundamental shift in how organizations understand and improve system performance.

THE CRITICAL DISTINCTION

Traditional process management focuses on defining workflows—how work is supposed to happen.

Process Intelligence focuses on understanding workflows—how work actually happens.

A defined process is static. A real system is dynamic.

Core Principles

- Adaptability over rigidity
- Flow over control
- Insight over assumption
- Enablement over enforcement

CASE — THE INVISIBLE BOTTLENECK

A product organization defines a clear delivery process: Design → Develop → Test → Deploy

On paper, the process appears efficient. In practice, work stalls between development and testing due to environment constraints. Testing teams become overwhelmed, creating delays. Developers continue producing work, increasing backlog.

Leadership sees increased activity—but not improved delivery. The issue is not the process definition. The issue is the lack of visibility into how the system actually behaves.

Once this flow is mapped, the bottleneck becomes clear. And once visible, it can be addressed. This is the power of Process Intelligence.

The Three Dimensions of Process Intelligence

At its core, Process Intelligence is not a tool or methodology—it is a system-level operating layer that enables organizations to understand, align, and continuously improve how work flows.

DIMENSION 1 — VISIBILITY

The ability to see how work actually flows across the system—not how it is assumed to flow.

False Visibility

Sprint progress. Ticket status. Deployment frequency. These provide snapshots of activity. They do not reveal system flow.

True Visibility Reveals:

- Where work slows
- Where queues form
- Where dependencies create delay
- Where handoffs introduce friction

Visibility does not improve the system directly. It makes improvement possible.

DIMENSION 2 — ALIGNMENT

The synchronization of teams, processes, and technology to operate cohesively rather than independently.

True Alignment Ensures:

- Work moves without friction between teams
- Ownership is clear across boundaries
- Priorities are consistent across functions
- Decisions reinforce system flow

Alignment does not increase speed. It removes resistance.

D I M E N S I O N 3 — O P T I M I Z A T I O N

The continuous improvement of system performance through targeted, data-informed adjustments.

Local Optimization

- Improve team performance
- Increase output
- Accelerate tasks

System Optimization

- Improve flow
- Reduce constraints
- Increase outcome reliability

Effective optimization focuses on the system constraint. Improving anything else produces minimal impact.

The Continuous Loop

Visibility → Alignment → Optimization → Repeat

This is not a sequence. It is a system.

Without visibility, alignment is based on assumptions.

Without alignment, optimization creates local improvements but global inefficiencies.

Without optimization, systems stagnate.

When all three operate together, the system becomes self-correcting, self-improving, and self-sustaining.

The Language of Process Intelligence

Instead of asking... **Ask instead...**

"Are we delivering faster?"	"Is the system flowing?"
"Are teams performing?"	"Is the system aligned?"
"Are we improving productivity?"	"Are we improving system performance?"

Language shapes perception. Perception shapes decisions. Process Intelligence begins with how we describe the system.

CHAPTER SYNTHESIS

Process Intelligence is not another delivery methodology. It is a shift in how systems are understood.

When visibility, alignment, and optimization operate together, improvement becomes continuous rather than reactive. This changes the role of process itself.

Process is no longer treated as rigid control. It becomes a living system of operational awareness.

DEEPENING THE PROCESS INTELLIGENCE MODEL™

- The Process Intelligence Model™ introduces three core dimensions:
- At a high level, this appears simple.
- In practice, each dimension represents a complex system capability that most organizations misunderstand.
- The purpose of this section is not to redefine the model—but to deepen its application.

VISIBILITY -SEEING THE SYSTEM AS IT IS

- Visibility is not reporting.
- It is not dashboards.
- It is not status updates.
- True visibility answers a different question:
- How does work actually move across the system?

FALSE VISIBILITY

- Most organizations believe they have visibility because they track:
- Sprint progress

- Ticket status
- Deployment frequency
- These provide snapshots of activity.
- They do not reveal system flow.

TRUE VISIBILITY

- True visibility reveals:
- Where work slows
- Where queues form
- Where dependencies create delay
- Where handoffs introduce friction
- It exposes reality—not intention.

FAILURE MODE

- Organizations operate based on assumed workflows.
- Decisions are made based on what should be happening rather than what is happening.
- This creates blind optimization.

SYSTEM SHIFT

- When visibility is achieved:
- Bottlenecks become obvious
- Delays become explainable
- System behavior becomes predictable
- Visibility does not improve the system directly.
- It makes improvement possible.

ALIGNMENT -SYNCHRONIZING THE SYSTEM

- Alignment is often misunderstood as communication.
- It is not.
- Alignment is system synchronization.

MISALIGNMENT IN PRACTICE

- Common patterns:
- Teams optimize for different metrics
- Handoffs lack ownership
- Priorities shift across functions
- Local success creates global failure
- Each team appears effective.
- The system does not.

TRUE ALIGNMENT

- Alignment ensures:
- Work moves without friction between teams

- Ownership is clear across boundaries
- Priorities are consistent across functions
- Decisions reinforce system flow

FAILURE MODE

- Organizations scale effort within misaligned systems.
- This increases:
- Friction
- Delays
- Complexity

SYSTEM SHIFT

- When alignment is achieved:
- Work flows predictably
- Teams operate with shared intent
- System performance stabilizes
- Alignment does not increase speed.
- It removes resistance.

OPTIMIZATION — IMPROVING SYSTEM PERFORMANCE

- Optimization is not about doing more.
- It is about improving flow.

LOCAL VS SYSTEM OPTIMIZATION

Local optimization:

- Improve team performance
- Increase output
- Accelerate tasks
- System optimization:
- Improve flow
- Reduce constraints
- Increase outcome reliability

FAILURE MODE

- Organizations optimize everywhere.
- This diffuses effort and reduces impact.

TRUE OPTIMIZATION

- Effective optimization focuses on:
- The system constraint
- Improving anything else produces minimal impact.

SYSTEM SHIFT

- When optimization is applied correctly:
- Small changes create large improvements
- Effort is focused
- Performance becomes sustainable

THE CONTINUOUS LOOP

- Visibility → Alignment → Optimization → Repeat
- This is not a sequence.
- It is a system.
- When continuously applied:
- Systems self-correct
- Performance stabilizes
- Improvement compounds
- Delayed decision cycles

CHAPTER 5

THE PROCESS ENABLEMENT GUILD

As organizations evolve, the role of process within the system must also evolve. Traditional process teams were built around control. Their primary responsibility was to define workflows, enforce compliance, and ensure adherence to established standards.

Modern systems require something different. They require enablement.

The Process Enablement Guild represents this shift. Rather than enforcing how work should be done, it focuses on improving how work actually flows.

THE GUILD'S ROLE

It operates as a system-level function—observing, aligning, and removing friction across teams, processes, and technology.

Its role is not to control. Its role is to enable performance.

Real-World Scenario

In a large organization, a centralized process team is responsible for defining development workflows. Teams are required to follow strict guidelines for planning, execution, and reporting. While this creates consistency, it also introduces rigidity.

Teams spend time complying with process requirements rather than improving delivery outcomes. When challenges arise, teams work around the process instead of through it. The system becomes fragmented.

Now consider a different approach. The organization introduces a Process Enablement Guild. Instead of enforcing compliance, the guild works with teams to:

- Identify bottlenecks
- Improve handoffs

- Align workflows across functions
- Remove unnecessary constraints

Over time, the system becomes more cohesive. Teams operate with greater clarity and less friction. The focus shifts from compliance to performance.

EXECUTIVE INSIGHT

Alignment—not effort—drives performance. Increasing effort within a misaligned system does not improve outcomes. It increases strain. True performance improvement comes from aligning how work flows across the system.

What the Guild Actually Does

GUILD FUNCTIONS

Visibility

Provides system-level insight into how work flows across teams.

Alignment

Ensures that teams operate with shared understanding and coordinated execution.

Optimization

Identifies and addresses constraints that limit system performance.

Common Failure Mode

Organizations attempt to modernize process by renaming existing functions without changing their behavior. A "process improvement team" becomes an "enablement team"—but continues to enforce compliance.

The shift to enablement requires a change in mindset:

- From control to collaboration
- From enforcement to insight
- From process ownership to system stewardship

CHAPTER SYNTHESIS

The Process Enablement Guild represents a structural shift in how organizations approach improvement.

It moves process from a function of control to a function of system optimization. By focusing on visibility, alignment, and continuous improvement, it enables the system to perform at a higher level.

CHAPTER 6

CONTINUOUS VALUE DELIVERY

Modern organizations are no longer defined by isolated releases. They are defined by continuous delivery of value. Value does not emerge from individual activities. It emerges from how work flows through the system.

Traditional models viewed delivery as a sequence of phases:

Plan → Build → Test → Release

Each phase was treated as a separate function. This created delays, handoffs, and fragmentation.

Continuous Value Delivery reframes this model. It treats delivery as an integrated system:

Plan → Build → Test → Release → Monitor → Learn → Repeat

Real-World Scenario

A product team operates on a two-week sprint cycle. At the end of each sprint, features are completed—but not released. They wait for a scheduled release window. During this time, additional changes accumulate. Testing becomes more complex. Risk increases. Delays occur.

THE ALTERNATIVE

Now consider a system designed for continuous flow. Work is released as soon as it is ready. Feedback is gathered immediately. Adjustments are made in real time.

The system becomes responsive, adaptive, and efficient. The difference is not effort. It is design.

EXECUTIVE INSIGHT

Sustainable speed is the result of system design—not effort. When systems are aligned, speed emerges naturally. When systems are misaligned, speed must be forced—and cannot be sustained.

What Breaks Continuous Flow

FLOW DISRUPTORS

Batching

Large releases increase complexity and risk.

Delayed Feedback

Issues are identified too late to be addressed efficiently.

Misaligned Incentives

Teams optimize for different outcomes.

Overloaded Systems

Too much work in progress reduces flow efficiency.

CHAPTER SYNTHESIS

Continuous Value Delivery is not achieved by accelerating individual activities. It is achieved by designing systems that enable flow.

When work moves smoothly across the system, value is delivered consistently and predictably.

C H A P T E R 7

PROCESS INTELLIGENCE IN PRACTICE

Understanding a system is one challenge. Applying that understanding consistently across an organization is another. Many organizations recognize inefficiencies, bottlenecks, and misalignment—but struggle to translate insight into sustainable improvement.

The purpose of Process Intelligence in practice is not to impose another transformation initiative. It is to establish a continuous system of visibility, alignment, and optimization that evolves alongside the organization itself.

Week 1 — Establish Visibility

FIRST PRIORITY: UNDERSTAND

The first step is not optimization. It is understanding. Most organizations attempt to improve systems before clearly seeing how work actually flows.

- Mapping how work moves across teams
- Identifying handoff points
- Detecting delays and queue accumulation
- Understanding where decision-making slows flow
- Identifying hidden dependencies

The objective is to reveal operational reality—not to document an idealized process.

Week 2 — Align the System

Once visibility is established, patterns of misalignment become easier to identify. Common misalignment patterns include:

- Teams optimizing for conflicting metrics
- Delivery schedules disconnected from operational readiness
- Unclear ownership between functional groups
- Priorities shifting faster than systems can adapt

Alignment requires more than communication. It requires system coherence. At this stage, organizations begin:

- Aligning metrics across functions
- Clarifying ownership boundaries
- Reducing friction between handoffs
- Standardizing operational expectations

Week 3 — Apply Targeted Optimization

CRITICAL SEQUENCE

Only after visibility and alignment are established does optimization become effective.

Most organizations reverse this sequence. They optimize first. As a result, they accelerate instability.

Rather than attempting broad transformation, focus on the primary system constraint. The key question becomes:

"What is limiting overall system flow?"

Week 4 — Measure System Impact

Improvement must be validated at the system level. Process Intelligence focuses on measuring outcomes such as:

- Lead time reduction

- Improved deployment reliability
- Reduced failure rates
- Increased delivery predictability
- Reduced operational friction
- Faster feedback cycles

EXECUTIVE INSIGHT

Metrics are not used to evaluate individuals. They are used to understand systems. When metrics are used for blame, teams protect themselves. When metrics are used for understanding, teams improve collaboratively.

CHAPTER SYNTHESIS

Process Intelligence in practice is not about enforcing rigid transformation programs. It is about establishing continuous operational awareness.

Organizations that apply Process Intelligence successfully do not merely move faster. They develop the ability to continuously understand, adapt, and improve how work flows across the system.

CHAPTER 8

THE IPI LIFECYCLE

FIGURE 5

Source: Intelligent Systems Press

Improvement is not a one-time event. It is a continuous process. Organizations often attempt large-scale transformations—restructuring teams, introducing new tools, or redefining processes—in the hope of achieving breakthrough results.

These efforts are costly, disruptive, and often short-lived. Sustainable improvement occurs through continuous, system-level refinement. The IPI Lifecycle provides a structured approach to achieving this.

The Six Stages

STAGE 1 — DISCOVER

Improvement begins with visibility. Organizations must first understand how work actually flows across the system.

- Bottlenecks
- Delays
- Dependencies
- Variability in performance

Without discovery, improvement efforts are based on assumptions. Assumptions lead to ineffective solutions.

STAGE 2 — PRIORITIZE

Not all issues are equal. Some constraints have a greater impact on system performance than others. Prioritization ensures that effort is focused where it matters most.

The goal is not to fix everything. The goal is to address the primary constraint limiting system performance.

STAGE 3 — EXPERIMENT

Once a constraint is identified, the next step is to test potential improvements. Experimentation should be controlled, measurable, and iterative. Small changes are introduced and evaluated.

Organizations that skip experimentation often implement large changes that fail to deliver expected results.

STAGE 4 — AUTOMATE

When an improvement proves effective, it should be scaled. Automation enables consistency and efficiency.

However, automation should only be applied after understanding the system. Automating an inefficient process accelerates inefficiency. This is a common failure point.

STAGE 5 — MEASURE

Improvement must be validated. Key metrics include:

- Lead time
- Cycle time
- Failure rate
- Throughput

Without measurement, improvement cannot be sustained.

STAGE 6 — EVOLVE

Systems are dynamic. What works today may not work tomorrow. Insights from measurement inform the next cycle of discovery. This creates a continuous loop of improvement.

CASE — FOCUSED IMPROVEMENT

An organization identifies multiple areas of inefficiency—testing delays, deployment bottlenecks, and unclear requirements. Rather than attempting to address all issues simultaneously, the team focuses on the most impactful constraint.

Testing delays are identified as the primary bottleneck. By addressing this constraint first, overall system performance improves significantly.

Improvement is most effective when it is focused.

CHAPTER SYNTHESIS

The IPI Lifecycle transforms improvement from a reactive activity into a continuous system. It provides a structured yet flexible approach to understanding, refining, and optimizing system performance.

The key is not to pursue perfection. The key is to pursue continuous progress.

CHAPTER 9

METRICS THAT MATTER

Measurement is one of the most misunderstood aspects of system improvement. Organizations often collect large amounts of data—yet struggle to derive meaningful insight.

THE CORE PROBLEM

The problem is not the lack of metrics. It is the selection of the wrong metrics.

Metrics shape behavior. What is measured becomes the focus of attention. If metrics are misaligned with system outcomes, behavior becomes misaligned as well.

Process Intelligence emphasizes measuring outcomes—not activity.

Activity vs. Outcome Metrics

Activity Metrics

Track what teams do

- Number of tasks completed
- Hours worked
- Story points delivered

These can indicate effort—but not effectiveness.

Outcome Metrics

Track what the system achieves

- Lead time
- Cycle time
- Deployment frequency
- Failure rate
- Mean Time to Recovery (MTTR)

Key Metrics for System Performance

Lead Time

The total time from when work begins to when it is delivered.

Cycle Time

The time required to complete a specific unit of work.

Deployment Frequency

How often new work is delivered to production.

Failure Rate

The percentage of changes that result in failure.

Mean Time to Recovery (MTTR)

The time required to recover from a failure.

CASE — THE METRIC FAILURE

A team tracks the number of features delivered per sprint. Performance appears strong. However, users report delays in receiving updates, and production issues increase.

When system-level metrics are introduced, a different picture emerges: Lead time is high. Failure rates are increasing. Recovery times are slow.

The issue is not output. It is system performance. By shifting focus to outcome metrics, the organization identifies underlying inefficiencies and begins to address them.

EXECUTIVE INSIGHT

What you measure determines what improves. If you measure activity, you improve activity. If you measure outcomes, you improve performance.

Common Metric Failures

PITFALLS TO AVOID

Vanity Metrics

Metrics that appear impressive but do not reflect meaningful outcomes.

Over-Monitoring

Tracking too many metrics without clear purpose.

Lagging Indicators Only

Focusing only on past performance without understanding current system behavior.

Lack of Context

Interpreting metrics in isolation rather than as part of a system.

CHAPTER SYNTHESIS

Metrics influence behavior. What organizations choose to measure ultimately shapes how systems operate.

Process Intelligence reframes measurement around flow, stability, adaptability, and system effectiveness.

The objective is not to generate more reporting. The objective is to generate clearer understanding.

CHAPTER 10

SYSTEM SIGNALS VS SYSTEM NOISE

Modern systems generate enormous amounts of information. Dashboards expand. Alerts multiply. Reports accumulate. Metrics increase continuously.

Yet despite this abundance of data, many organizations struggle to understand what is actually happening inside their systems.

THE CORE PROBLEM

The problem is not lack of information. The problem is the inability to distinguish meaningful system signals from operational noise.

Because systems overwhelmed by noise lose the ability to respond intelligently.

What Is a System Signal?

A system signal reveals meaningful insight into system behavior. It identifies conditions that directly influence flow, stability, alignment, or performance.

EXAMPLES OF MEANINGFUL SIGNALS

- Increasing lead time
- Rising recovery duration
- Recurring deployment instability
- Persistent bottlenecks
- Escalating context switching
- Delayed feedback cycles
- Dependency accumulation

Signals reveal patterns. Patterns reveal system conditions. And system conditions determine performance.

What Is System Noise?

Noise is activity that appears important but provides little actionable understanding.

EXAMPLES OF NOISE

- Vanity metrics
- Excessive dashboard reporting
- Alert overload
- Isolated activity spikes
- Metrics without operational context
- Reporting disconnected from outcomes

Noise consumes attention without improving awareness. Over time, organizations become highly informed—but operationally blind.

The Consequence of Noise

Failure Pattern	Impact on System
Decision fatigue	Teams overloaded by excessive information
Delayed response	Critical signals become difficult to identify
Misaligned priorities	Attention shifts toward activity vs outcomes
Reactive operations	Responding to symptoms, not conditions

EXECUTIVE INSIGHT

The objective is not maximum visibility. The objective is meaningful visibility. Too little visibility creates blindness. Too much visibility creates noise. Process Intelligence creates discernment between the two.

CHAPTER SYNTHESIS

Modern systems fail not because information is unavailable—but because meaningful signals are buried beneath operational noise.

Process Intelligence enables organizations to distinguish what matters from what distracts. This transforms visibility from passive reporting into actionable system awareness.

CHAPTER 11

THE PROCESS INTELLIGENCE MATURITY MODEL

Not all organizations operate at the same level of system awareness. Some operate reactively. Others achieve coordinated flow. A small number evolve into continuously adaptive systems.

The Process Intelligence Maturity Model provides a framework for understanding this progression. Its purpose is not classification. Its purpose is visibility.

LEVEL 1 — REACTIVE SYSTEMS

Organizations operate primarily through urgency and reaction.

- Frequent operational firefighting
- Delayed issue detection
- Fragmented communication
- Heavy dependency on individual effort
- Inconsistent delivery outcomes

Primary Constraint:

Lack of visibility into system behavior.

LEVEL 2 — VISIBLE TEAMS

Organizations begin improving team-level processes. Visibility improves within teams—but remains limited across the broader system.

- Improved sprint execution
- Better team collaboration
- Local process optimization
- Persistent cross-functional friction

Primary Constraint:
Lack of cross-system alignment.

LEVEL 3 — ALIGNED DELIVERY

Organizations begin coordinating flow across functions. Dependencies become more visible.

- Shared delivery objectives
- Improved cross-functional coordination
- Reduced handoff friction
- Increased delivery predictability

Primary Constraint:
Limited continuous optimization capability.

LEVEL 4 — OPTIMIZED FLOW

Organizations operate with strong visibility, alignment, and targeted optimization practices.

- Continuous delivery optimization
- Constraint-focused improvement
- High operational predictability
- Data-informed decision-making

Primary Constraint:
Scalability of intelligence across complex systems.

LEVEL 5 — INTELLIGENT ADAPTIVE SYSTEMS

At the highest level, organizations evolve into continuously learning systems. Visibility, alignment, optimization, automation, and feedback operate together dynamically.

- Real-time operational insight
- Adaptive optimization
- AI-assisted system intelligence
- Continuous organizational learning
- Predictive system management

The organization no longer relies primarily on reactive intervention. The system continuously adjusts itself through visibility and feedback.

EXECUTIVE INSIGHT

Advancing maturity is not about implementing more tools. It is about increasing system awareness. Maturity develops progressively: Visibility enables alignment. Alignment enables optimization. Optimization enables intelligent adaptation.

CHAPTER SYNTHESIS

System maturity is not determined by the number of tools an organization adopts. It is determined by the quality of operational awareness the organization develops over time.

Maturity is therefore not a destination. It is the continuous evolution of system understanding.

CHAPTER 12

ENGINEERING SYSTEM BALANCE

FIGURE 7

SYSTEM BALANCE TRIANGLE

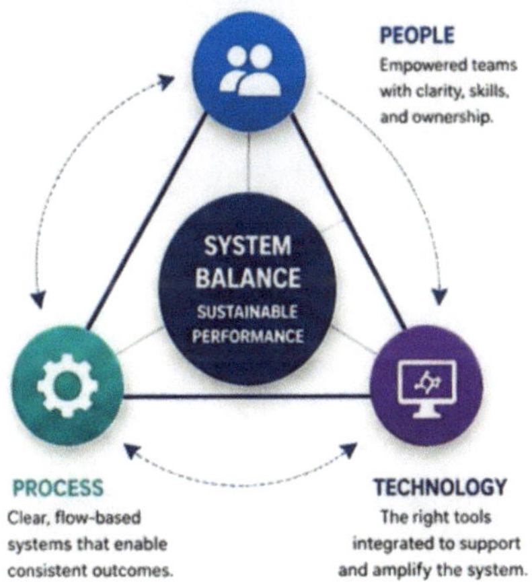

Source: Intelligent Systems Press

System performance is not determined by any single component. It is determined by balance. Organizations often focus on optimizing individual elements—people, processes, or technology.

However, when these elements are not aligned, improvements in one area create instability in another.

This concept builds on ideas explored in Feng Shui for Intelligent Process Improvement, where system harmony was identified as a key driver of sustainable performance.

THREE CORE DIMENSIONS

People

Execute the work.

Process

Guide the work.

Technology

Enable the work.

When these elements are aligned, systems operate efficiently. When they are not, friction emerges.

Real-World Scenario

An organization invests heavily in automation tools to improve delivery speed. Technology capabilities increase significantly. However, teams are not trained to use these tools effectively, and existing processes are not adapted to support them.

The result: underutilized tools, increased complexity, reduced efficiency. The system becomes imbalanced. The issue is not technology. It is alignment.

The Consequences of Imbalance

IMBALANCE PATTERNS

Overemphasis on People

Relies on effort rather than system design, leading to burnout.

Overemphasis on Process

Creates rigidity, reducing adaptability.

Overemphasis on Technology
Introduces complexity without improving flow.

Balanced systems distribute capability across all three dimensions.

EXECUTIVE INSIGHT
Systems fail quietly before they fail visibly. Imbalance often develops gradually. Performance appears stable—until it is not. Process Intelligence provides the visibility needed to detect and correct imbalance early.

Designing for Balance

- Align processes with team capabilities
- Ensure technology supports—not dictates—workflow
- Continuously evaluate system performance across all dimensions

Balance is not a static state. It is continuously maintained through observation and adjustment.

CHAPTER SYNTHESIS
System balance is essential for sustainable performance. Optimizing individual components is insufficient.

Only when people, process, and technology operate in alignment can systems achieve consistent, high-level performance.

CHAPTER 13

THE COST OF INVISIBLE WORK

Not all work inside a system is visible. In many organizations, some of the most expensive operational activity never appears in dashboards, reports, or delivery metrics.

WHAT IS INVISIBLE WORK?

- Context switching
- Rework
- Waiting for approvals
- Clarifying unclear requirements
- Managing fragmented communication
- Resolving dependency conflicts
- Recovering from preventable failures

These activities consume time, attention, and cognitive capacity. Yet they are rarely measured directly. As a result, organizations underestimate the true cost of system inefficiency.

The Illusion of Productivity

Many organizations evaluate productivity based on visible activity: tasks completed, features delivered, hours worked, sprint velocity.

However, these indicators often conceal operational friction occurring beneath the surface. A team may appear highly productive while simultaneously experiencing:

- Excessive interruptions
- Constant priority shifts
- Unstable workflows
- High mental fragmentation

- Continuous operational recovery

The result is exhaustion without sustainable progress.

The Cost of Context Switching

Modern systems frequently overload teams with simultaneous demands. Engineers shift continuously between development tasks, production incidents, meetings, Slack messages, approval requests, and tooling interruptions.

> Each interruption imposes a cognitive reset cost. Over time, this fragmentation significantly reduces deep focus and system effectiveness.
>
> The issue is not lack of effort. It is degraded flow.

The Human Impact of Invisible Work

Invisible work affects more than operational efficiency. It affects people. When systems remain fragmented:

- Cognitive fatigue increases
- Burnout accelerates
- Motivation declines
- Attention quality deteriorates
- Collaboration weakens

Organizations often interpret these symptoms as personnel problems. In reality, they are frequently system problems.

EXECUTIVE INSIGHT

Invisible work is not free. Organizations pay for it through delays, burnout, turnover, reduced innovation, operational instability, and lost strategic focus.

CHAPTER SYNTHESIS

The most damaging inefficiencies inside modern systems are often the least visible. Not because they are insignificant—but because organizations have learned to normalize them.

Process Intelligence reveals these conditions not to increase control—but to reduce unnecessary friction and restore sustainable flow.

Because systems perform best when both operational efficiency and human capacity are protected simultaneously.

CHAPTER 14

AI AND PROCESS ENGINEERING

FIGURE 8

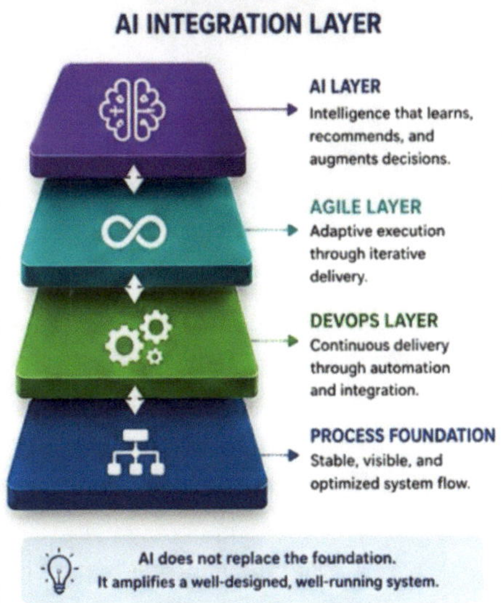

Source: Intelligent Systems Press

Artificial Intelligence represents the next major evolution in software delivery. It introduces the ability to automate decision-making, accelerate execution, and generate insights at scale.

THE CRITICAL PRINCIPLE

AI does not fix broken systems. It amplifies them.

If processes, workflows, and decision-making structures are aligned, AI increases efficiency and performance.

If they are not, AI increases complexity and risk.

CASE — AI WITHOUT SYSTEM UNDERSTANDING

An organization integrates AI-driven testing tools to accelerate quality assurance. Test execution becomes faster. Coverage improves. However, underlying issues in requirements clarity and system alignment remain.

Defects continue to emerge. AI improves execution—but does not address root causes.

AI enhances systems—but does not define them.

The Role of AI in Process Intelligence

Within the Process Intelligence framework, AI serves as an enabler. It supports:

- Pattern recognition across workflows
- Identification of bottlenecks and anomalies
- Predictive insights for system optimization
- Automation of repetitive tasks

However, these capabilities are only effective when applied within a well-understood system. Without visibility and alignment, AI operates in isolation.

EXECUTIVE INSIGHT

The future is AI within well-designed systems. Organizations that prioritize system design will maximize the value of AI. Those that rely on AI to compensate for system deficiencies will encounter increasing complexity. AI is not a replacement for Process Intelligence. It is an extension of it.

Executive Decision Framework

At the executive level, the challenge is not managing tasks—it is understanding systems. Process Intelligence shifts this focus to system alignment.

Wrong Response	Correct Response
When delivery slows: Ask which team is underperforming	Ask where flow is breaking down
When failures increase: Ask who made the mistake	Ask which system condition enabled the failure
When output increases but outcomes do not: Push for more effort	Examine system alignment

CHAPTER SYNTHESIS

AI increases the speed, scale, and adaptability of modern systems. But acceleration alone does not create intelligence.

Process Intelligence provides the visibility and feedback necessary to ensure intelligent systems remain understandable, trustworthy, and adaptable as complexity continues to increase.

CHAPTER 15

AI GOVERNANCE WITHIN PROCESS INTELLIGENCE

As AI capabilities expand, organizations face a new challenge: not simply how to use AI—but how to govern it responsibly within complex systems.

Without governance, AI can amplify instability as quickly as it amplifies efficiency. This is why Process Intelligence becomes increasingly important in AI-enabled environments.

The Governance Challenge

Traditional governance models were designed around predictable workflows and human-controlled decision-making. AI changes this dynamic.

CRITICAL CHALLENGES

- Reduced decision transparency
- Unclear accountability boundaries
- Escalation of automation errors
- Over-reliance on algorithmic output
- Feedback loops that reinforce flawed behavior

Without visibility into how AI influences system behavior, organizations risk accelerating hidden instability.

Human Oversight Remains Essential

AI should enhance system intelligence—not replace human judgment. Process Intelligence reinforces the importance of maintaining:

- Human validation of critical decisions
- Transparent feedback loops
- Clear accountability structures
- Continuous monitoring of system outcomes
- Operational visibility into AI behavior

The objective is not unrestricted automation. The objective is trustworthy system augmentation.

AI and System Drift

THE RISK OF DRIFT

As systems evolve, AI models may begin operating under conditions different from those originally intended. This creates gradual divergence between expected outcomes, actual behavior, and operational reality.

Because this drift often occurs incrementally, organizations may not recognize degradation until performance has already declined.

Process Intelligence helps identify this drift by continuously observing system behavior rather than assuming ongoing alignment.

Trust, Validation, and Adaptation

AI-enabled systems must be continuously validated. This includes:

- Verifying outcome quality
- Monitoring unintended consequences
- Measuring operational impact
- Reassessing alignment with organizational goals

Trust in AI should never be assumed. It should be continuously earned through visibility, validation, and feedback.

EXECUTIVE INSIGHT

The future will not belong to organizations with the most AI. It will belong to organizations that can govern intelligent systems responsibly. AI without governance creates fragility. AI within Process Intelligence creates adaptive capability.

CHAPTER SYNTHESIS

AI introduces extraordinary potential—but also increased system complexity.

Process Intelligence provides the operational visibility, alignment, and feedback mechanisms required to ensure AI enhances systems responsibly and sustainably.

The goal is not automation alone. The goal is intelligent, trustworthy system evolution.

C H A P T E R 1 6

PROCESS INTELLIGENCE DIAGNOSTIC

Use this diagnostic to assess your organization's current state of Process Intelligence maturity. This is not an evaluation tool. It is a visibility tool.

V I S I B I L I T Y

- Can you see real workflow across teams?
- Where does work slow consistently?
- Are bottlenecks visible or assumed?

A L I G N M E N T

- Are teams optimizing for shared outcomes?
- Are handoffs clearly owned?
- Do priorities conflict across functions?

O P T I M I Z A T I O N

- Is improvement focused on constraints?
- Are changes measured for impact?
- Is optimization continuous or reactive?

Scoring Your System

I N T E R P R E T A T I O N G U I D E

Mostly "No"

Fragmented system — significant opportunity to establish foundational visibility and alignment.

Mixed Results

Transitional system — visibility is improving but alignment gaps remain. Focus on cross-functional coordination.

Mostly "Yes"

Aligned system — foundation is strong. Focus on continuous optimization and scaling system intelligence.

This diagnostic reveals where the system needs attention. Use it as a starting point for continuous system improvement, not as a final assessment.

CHAPTER 17

EXECUTIVE SCENARIO PLAYBOOK

The following playbook provides executive-level guidance for responding to common system challenges through the lens of Process Intelligence.

WHEN DELIVERY SLOWS

✗ **Wrong Response:**

Increase pressure on teams

✓ **Correct Response:**

Identify where flow breaks

WHEN FAILURE RATES INCREASE

✗ **Wrong Response:**

Blame execution

✓ **Correct Response:**

Examine system conditions

WHEN OUTPUT INCREASES BUT OUTCOMES DO NOT

✗ **Wrong Response:**

Push for more effort

✓ **Correct Response:**

Assess system alignment

WHEN AI IS INTRODUCED

✗ **Wrong Response:**

Expect improvement

✓ **Correct Response:**

Validate system readiness

WHEN SCALING BREAKS

✗ **Wrong Response:**

Add resources

✓ **Correct Response:**

Identify system constraint

EXECUTIVE INSIGHT

The difference between managing operations and leading transformation lies in the questions asked. Systems thinking begins at the executive level.

CHAPTER 18

CASE STUDY

THE MISALIGNED PIPELINE

Organization: Mid-Sized SaaS Platform | Industry: Enterprise Software Delivery

The Illusion of Progress

The organization had everything. Agile teams were running efficiently. DevOps pipelines were fully automated. AI-assisted testing had been introduced. Velocity was increasing. Deployments were frequent.

ON DASHBOARDS — PERFORMANCE LOOKED STRONG

But in reality:

- Releases were delayed
- Failures were increasing
- Customers were reporting instability

The system was moving faster—but breaking more often.

The Real Problem

Leadership initially believed the issue was capacity. More engineers were hired. More automation was introduced. Nothing improved. Because the problem was not effort. It was misalignment across the system.

Local Optimization	**System Reality**
Development output increased	Work moved quickly within teams

QA struggled with overflow	But slowed dramatically between them
Operations absorbed instability	The system was misaligned

The Breakthrough Moment

THE SHIFT IN THINKING

They stopped asking: "Which team is underperforming?"

And started asking: "Where does the system break?"

They mapped the full system. For the first time, they saw:

- Bottlenecks at testing environments
- Delays at handoff points
- Misaligned incentives across teams

The constraint was not inside a team. It was between them.

The Transformation

PROCESS INTELLIGENCE APPLIED

1. Visibility

Mapped real system flow—not assumed workflows

2. Alignment

Redefined ownership across the system—not within silos

3. Optimization

Targeted the constraint—not the symptoms

The Results

WITHIN MONTHS:

- Deployment frequency increased 3x
- Failure rates dropped significantly
- System stability improved
- Teams became aligned around shared outcomes

But the most important outcome was this:

They could now see their system. And once seen—it could be improved continuously.

CHAPTER 19

ABOUT THE AUTHORS

V. Quan Lee and J. Kay are practitioners and thought leaders in process engineering, system alignment, and modern delivery optimization, with nearly six decades of combined experience in Process Engineering.

Their work spans the evolution of process thinking—from requirements discovery and structured methodologies to system balance and intelligent, adaptive delivery systems. They have worked across complex organizational environments, focusing on how to improve the flow of work, eliminate inefficiencies, and align systems for continuous performance.

Published Works

BOOKS BY V. QUAN LEE & J. KAY

- Project Requirements Analysis: Book of Questions
- 12 Steps to Process Improvement
- Feng Shui for Intelligent Process Improvement
- Process Intelligence: The Missing Link Between Agile, DevOps, and AI

Through these works, they have developed a progressive body of thought that moves from questioning and clarity, to structure, to balance, and ultimately to Process Intelligence—a unified framework for improving how work flows across modern Agile, DevOps, and AI-driven environments.

EXECUTIVE INSIGHT

Their philosophy is grounded in a simple principle: Systems perform best when they are clearly understood, intentionally designed, and continuously improved.

C H A P T E R 2 0

ABOUT THE PUBLISHER

INTELLIGENT SYSTEMS PRESS

Intelligent Systems Press is an independent publishing imprint dedicated to advancing modern thinking in system design, process optimization, and intelligent delivery.

The imprint was established to explore how organizations can move beyond fragmented methodologies and toward integrated systems that continuously improve.

Its work sits at the intersection of structure, flow, and intelligence—bringing together principles from traditional process management, systems thinking, and emerging technologies.

Through its publications, Intelligent Systems Press examines how work actually moves across complex environments, emphasizing visibility, alignment, and the reduction of friction across people, processes, and technology.

Titles published under Intelligent Systems Press are designed to provide both conceptual clarity and practical application—bridging the gap between theory and execution.

Current Publications

- Project Requirements Analysis: Book of Questions
- 12 Steps to Process Improvement
- Feng Shui for Intelligent Process Improvement
- Process Intelligence: The Missing Link Between Agile, DevOps, and AI

Together, these works represent an evolving approach to modern process thinking—moving from requirements clarity, to structure, to balance, and ultimately to intelligent system optimization.

Intelligent Systems Press is committed to developing ideas that not only explain systems—but improve how they perform.

www.IntelligentSystemsPress.com

CHAPTER 21

YOUR FIRST 30 DAYS OF PROCESS INTELLIGENCE

Understanding a system intellectually is important. Learning to observe and improve that system operationally is where transformation begins.

The purpose of Process Intelligence is not to create another large-scale initiative. It is to develop continuous awareness of how work flows.

The following 30-day model provides a practical starting point for organizations beginning this shift.

WEEK 1 — OBSERVE THE SYSTEM

Do not optimize immediately. Observe first. Focus on understanding:

- Where work slows
- Where queues accumulate
- Where handoffs create friction
- Where communication breaks down
- Where dependencies delay flow

The objective is not judgment. The objective is visibility.

WEEK 2 — IDENTIFY THE PRIMARY CONSTRAINT

Every system has constraints. Some are visible. Others remain hidden beneath activity.

Identify the single condition creating the greatest limitation to system flow. Examples may include:

- Testing bottlenecks
- Approval delays
- Excessive work in progress
- Unclear ownership
- Fragmented priorities

Do not attempt to solve everything simultaneously. Focus creates leverage.

WEEK 3 — ALIGN THE SYSTEM

Once the constraint is visible, begin aligning the surrounding system. This may involve:

- Clarifying ownership
- Aligning metrics
- Simplifying handoffs
- Improving communication pathways
- Reducing unnecessary complexity

The objective is not control. The objective is coordinated flow.

WEEK 4 — APPLY TARGETED IMPROVEMENT

Introduce small, measurable improvements. Observe the effect on:

- Lead time
- Reliability
- Flow consistency
- Operational stability
- Team friction

Do not pursue perfection. Pursue understanding. The system will continue evolving. So must the approach used to improve it.

The Shift

Process Intelligence is not implemented through a single transformation program. It develops through continuous system awareness.

EXECUTIVE INSIGHT

Organizations that learn to see their systems clearly gain the ability to improve continuously, intentionally, and intelligently. This is where the shift begins.

CHAPTER 22

FINAL THOUGHT

Most organizations will continue to optimize parts.

A few will begin to understand systems.

The difference between the two defines the future of performance.

Process Intelligence is not about doing more. It is about seeing clearly.

Once a system is understood, it can be improved—continuously, intentionally, and intelligently.

The question is not whether this approach works.

The question is whether organizations are ready to see what has been there all along.

Because once it is seen—it cannot be unseen.

Start the Shift

If you are ready to move beyond fragmented improvement and into true system performance—start here.

APPLY PROCESS INTELLIGENCE

- Map your system

- Identify your constraint
- Align your teams
- Optimize continuously

Connect & Continue

For frameworks, tools, and insights:

Intelligent Systems Press

www.IntelligentSystemsPress.com

The future of operational performance will not be defined by isolated frameworks, faster tooling, or increased automation alone.

It will be defined by how clearly organizations understand the systems through which work actually flows.

Process Intelligence represents a shift toward that understanding. Not as a static methodology—but as a continuous system of visibility, alignment, adaptation, and improvement.

Because systems that can understand themselves can continuously evolve themselves.

THIS IS WHERE PROCESS INTELLIGENCE BEGINS.

www.ingramcontent.com/pod-product-compliance
Lightning Source LLC
LaVergne TN
LVHW051129210926
848482LV00001B/35